John A. Benson

Synopsis of a Course of Lectures

On the Anatomy, Physiology and Histo-Chemistry of the Nervous System

John A. Benson

Synopsis of a Course of Lectures
On the Anatomy, Physiology and Histo-Chemistry of the Nervous System

ISBN/EAN: 9783337140359

Printed in Europe, USA, Canada, Australia, Japan

Cover: Foto ©berggeist007 / pixelio.de

More available books at **www.hansebooks.com**

SYNOPSIS

OF A

COURSE OF LECTURES

ON THE

ANATOMY,

PHYSIOLOGY AND HISTO-CHEMISTRY

OF THE NERVOUS SYSTEM.

JOHN A. BENSON, M. D., COLUMB.

Professor of Physiology College of Physicians and Surgeons,

CHICAGO. ILL.

CHICAGO:

E. H. COLEGROVE & COMPANY,

1895

Respectfully Inscribed
to
Dr. Dan'l R. Brower.

PART I.

Biological Introduction.

The determining characteristics of living things. Barriers
  separating animate and inanimate worlds. The Cell—
  peri-nuclear portion, the nucleus, nucleolus, etc.
  Properties of Protoplasm. Observations of Authori-
  ties, to-wit: Schwann, Dujardin, etc., to the present
  time. Karyokinesis.

Unicellular plants—"Torula," "Protococcus pluvialis."

Unicellular animals—"Proteus animalcule."

Parasitic organisms—"Pencillium glaucum," "Mucor mu-
  cedo," "Bacteria."

Unicellular animals with differentiation of structure—
  "Vorticella."

Multicellular organisms—"Hydra viridis," "Hydra fusca."

Formation of a nervous system—Beginning of differentiation of nervous from other elements, as illustrated in
  simpler forms of life. Formation of reflex loop. Higher
  arrangement of ganglia and fibres. Ratio of fusion of
  nerve cells and strands to integration of function.
  Functions of the nervous system in associating different
  parts of the same body; in bringing the living being
  into relation with the external world.

Reproduction—Embryonic growth and development with
  special reference to the nervous system.

The law of periodicity, or rhythm in nature—Examples,
  etc.

The law of habit—Application, examples.

The origin of the forms of life—The doctrine of Evolution; progressive integration of structure and function,
  with passage from the single or uniform to the multiple

(4)

or multiform, from the simple to the complex; and from the general to the special. Physiological division of labor, or specialization of function and differentiation of structure.

Origin of life—Two principal views: "The theory of creation," and the "theory of descent with modification." Teachings of Charles Darwin. The survival of the fittest. Sexual selection.

Arrangement of evidence—1, Morphology; 2, Embryology; 3, Mimicry; 4, Rudimentary organs; 5, Geographical distribution; 6, Paleontology; 7, Fossil and existing species; 8, Progression; 9, Domesticated animals. (Mills).

The American school—Teachings of Cope and of Hyatt. Investigations and observations of Rev. Dr. J. A. Zahm. Neo-Lamarckianism. Weismann's teachings. Theories of Brooks. Darwin's hypothesis of "pangenesis." Vines' equivalent of "parthenogenesis." St. George Mivart's doctrine of "extraordinary births." Romane's idea of "physiological selection." Application to the consideration of diathesis, idiosyncrasy. Observations of Herbert Spencer.

Man's place in the animal kingdom—No longer in zoology placed in a separate group by himself, man now classed with the "primates" along with the anthropoid apes (gorilla, etc.), the simians of the old and new worlds and the lemurs. Structural resemblances between man and the other primates very great. More difference between the structure of the most widely separated members of the group than between certain of the anthropoid apes and man. Points of greatest resemblance between man and the anthropoid apes are: The same number of vertebræ; the same general shape of the pelvis; a brain distinguishing them from other mammals; and posture, being bipeds. Distinctive characters are: Size, rather than form of the brain,

that of man being more than twice as large; a relatively larger cranial base, by which, together with the greater size of the jaws, the face becomes prominent, the earlier closure of the sutures of the cranium arresting the growth of the brain; more developed canine teeth and difference in the order of eruption of the permanent teeth; the more posterior position of the foramen magnum; the relative length of the limbs to each other and to the rest of the body; minor differences in the hands and feet, especially the greater freedom and power of apposition of the great-toe. Greatest distinction between man and his closest allies among the apes is the development to a higher degree of his intellectual and moral nature, corresponding to the differences in weight and structure of the human brain, and associated with the use of spoken and written language; experience of previous generations, not only registered in the organism (heredity), but also in a form more quickly available (books, etc.). The greatest structural difference between the races of men are referable to the cranium, since all interbreed freely, must be considered varieties of one species. Ethnological and anthropological observations. Teachings of Wesley Mills, etc. Cuvier's law of the conservation of the species. Nature detests hybrids. Tendencies to pass from higher to lower planes.

Methods of Examination of Nervous Tissues.

A. *Defibering*—The methods of Ruysch and of Vicq d'Azyr. of Reil, of J. Stilling. Uncertainty of these plans.

B. *The preparation of section series*—Stilling's method. Apparatus of John C. Dalton. Hardening the tissues. Fixing media. Section cutting. Microtomes, various kinds. Gudden's, Reichert's sledge, Weigert's modification of the diving microtome of Schanze. Roy's

machine. Celloidin method. Staining methods of Gerlach, of Hoyer, of Loewenthal. *Nuclear stains*—Alum-Hæmatoxylin. Csoker's method. Grenacher's method. *Medullary sheath stains*—Exner's perosmic acid method. Palladium and gold. Weigert's and Pal's methods. *Axis cylinder stains*—Carmine. Freud's method. *Other methods of coloring. Impregnating with metals, etc.*—Golgi's sublimate coloring. Adamkiewicz' staining in saffronin. Glycerin mounting. Oblique illumination in microscopical view. Flesch's artifice in the use of colored lights.

C. *Investigation of central nervous system in embryonic and pathological conditions*—Flechsig's observations on nerve myelination. Secondary degeneration. Observations of Rokitansky and Tuerck. Wallerian law. Histogeny and nutrition of nerves. Gudden's method.

D. *Comparative method*—Relation of organic functional importance to anatomical prominence. Edinger's combination of comparative and historical methods. Meynert's observations.

E. *Physiological method*—Vivisection. Irritation and ablation of nerves ; of cord ; of brain, etc. Electrotonus. Experimental irritation and ablation compared with pathological processes, lesions, etc. Value of physiological experiments to be estimated only in careful association with anatomical data.

Histo-Chemistry of the Nervous System.

Classification of Nerve organs. Gray and white matters. Cells and fibers. Neurilemma. Axis cylinder. Schwann's medullary tube. Axilemma. Medulemma. *Proteids* of nervous tissues. Importance of thorough study of the proteids. Their peculiar place, and the necessity of our giving to them exceptional treatment, for they alone are never absent from the active living cells, which we

(7)

recognize as the primordial structures of animal and vegetable organisms. The material substratum of the animal organism is proteid, and it is through the agency of structures essentially proteid in nature that the chemical and mechanical processes of the body are effected. Proteids are indispensable constituents of every living, active, animal tissue, and are indissolubly connected with every manifestation of animal activity. *Neurokeratin.* Its properties and mode of separation. *Nuclein*, analysis and amount. *Phosphorized constituents of nervous tissues*—Protagon. Discovery by Liebrich; mode of preparation and properties. Hypotheses of Diaconow, of Hoppe-Seyler, of Thudicum. Method of Blankenhorn and Arthur Gamgee. Results of analysis. Stability. Researches on products of decomposition. Lecithin. Observations of Gobling, Hoppe-Seyler and Diaconow. Formation of compounds. How separated from brain tissue. Analysis. Products of decomposition, to wit: glycerin-phosphoric acid and neurin or cholin. *Other phosphorized bodies*—Kephalines, Myelins and Lecithins. *Non-phosphorized bodies of unknown constitution*—Cerebrin or cerebrins. Mueller's researches. Modes of preparation as used by Geoghagan of Dublin, and Gamgee. Pseudocerebrin. Researches as to its decomposition. Cetylid. Thudicum's researches. Cerebrin, phrenosine and kerasine. *Monad-alcohol*—Cholestrine, preparation, properties, compounds and derivatives. Observations of Austin Flint. Cholestræmia. *Extractives common to nervous and other tissues*—Organic and inorganic constituents. Comparative analysis of gray and white matters. *Review and examination* of the chemical processes connected with the activity and death of nervous tissues. Observations of Gscheidlen, Liebrich and Funke. *Chemical history* of certain of the peri-

(8)

pheral terminations of the nervous system and of the accessory structures connected with them, to wit: the tissues and media of the ear and eye. *Tissues and media of the ear*—Perilymph and endolymph. Otoliths, lapilli or otoconia. Membranous labyrinth, etc. *Tissues and media of the eye*—Cornea, analysis of. Presence of myosin. Sclerotic. Aqueous humor; physical properties and chemical constituents. Crystalline lens, analysis, chemical constituents, changes in cataract. Vitreous body and choroid. Retina, subdivision into layers. Particular description of the visual epithelium (retinal epithelium). Tapetum. Variations of retinal structures in different regions or areas and in different species of animals. ' General chemistry of the retina. Chemical structure of inner and outer limbs of rods and cones. Solubility of outer limbs in bile. Chromophanes or coloring matters of cones; chlorophanes, xanthophane, rhodophane. *Rhodopsin*, or visual purple of rods. *Historical*—Distribution in the retina. Mode of separation. Optical characters. Spectrum. Effects of light of different wave lengths. Influence of temperature. Action of re-agents. Effects of loss of water. Optograms, retinal photography. Chemical facts. Myeloidin. Fat. Lipochrin. Fuscin. Action of light on visual purple in living eye. Visual yellow. Regeneration. Vision without visual purple.

Morphology of the Central Nervous System.

Macroscopic demonstrations. Leuhossek's and Schwalbe's methods. Histogeny. Development of ganglionic rudiments. Neuroblasts and spongioblasts. Observations of His on embryonic growth. Neuroglia cells. Myelin cells. Development of spinal and cranial nerves. Development of the brain. The velum interpositum. The cerebral vesicles. Fore-brain, 'tween-

brain, mid-brain, hind-brain and after-brain. Development of cortex and stem. Encephalo-spinal axis. Intermediate gray matter of the crural system. Connections of the nucleus caudatus with the rind of the frontal and tempero-sphenoidal lobes. Formation of the ventricles.

General Anatomy and Consideration of the Intimate Structure of the Brain and Bulb.

1. *The after-brain*—Bulbar region.

2. *The hind-brain*—Surface, lobes, Flocculus, Vermis. The Medullary centre. Arbor-vitæ. Nucleus of the roof. Cerebellar peduncular systems. Floor of the fourth ventricle. Sinus Rhomboidalis. Vela medullaria. Foramen of Magendie.

3. *The mid-brain*—Sulcus lateralis mesencephali. Relations of great cerebral peduncles. Inter-peduncular space. Corpora quadrigemina. Brachia, etc. Tegment. Substantia nigra. Semilunar sections of crusta. Nerves in this region.

4. *The 'tween-brain*—Optic thalamus. Corpora geniculata. Optic tracts and corpora mamillaria. Pulvinar. Cavity of the 'tween-brain, or third ventricle. Aditus ad aquæductum sylvii. Recessus infundibuli. Striæ corneæ. Ansa intergenicularis. Tænia thalami. Ganglion habenulæ. Conarium. Posterior commissure. Structures in ventral aspect.

5. *The great brain*—The hemispheres. The brain-mantle. The basal ganglia. Caudate nucleus. Lentiform body Globus pallidus primus and secundus. The Putamen. External capsule. Claustrum. Capsula extrema. The thalami. Regio sub-thalamica. Centrum semi-ovale Vieussenii. Internal capsule. Corpus callosum. Septum pellucidum. Fifth ventricle. Pedunculus septi pellucidi. Fornix, fimbria, crura fornicis.

Gray matter. Bundle of Vicq d'Azyr. Verga's ventricle. Anterior commissure, torsion of fibres in its course. Base of the fore-brain in front of the optic chiasm. Lamina cribrosa. Olfactory tract. Gray floor commissure. Lamina terminalis. Sulcus medius substantiæ perforatæ anterioris.

6. *Ventricles of the great-brain*—How formed. Topography. Connections, relations, etc. Ontogenetic study of the ventricles. Their morphological relations to the central nervous system. Rima transversa. Tela choroidea superior. Plexus choroideus lateralis. The Glomus. Sulcus choroideus. Plexus choroideus medialis. *Third ventricle-* Two portions, horizontal and vertical. The aditus. Floor. Recessus chiasmatis. Foramen of Monro. Sulcus Monroi. The psalterium. Spatium supra-choroideum. Ventricle of Verga. Venæ striæ corneæ. *Lateral ventricles*—Cella media, roof, walls, floor. Anterior horn, walls, etc. Posterior horn, tapetum, forceps posterior corporis callosi. Calcar avis. Fasciculus longitudinalis inferior. Inferior horn. Hippocampal fissure. Nucleus amygdaleus. Subiculum cornu ammonis. Substantiæ reticularis Arnoldi. Fascia dentata. Fimbria. Pes hippocampi majoris. Eminentia collateralis Meckelii. Fissura collateralis. Fissura subiculi interna. Relations of gyrus fornicatus, the indusium griseum, nervi Lancisii. Tuberculum fasciæ dentatæ. The callosal gyri.

7. *Topography of the brain-mantle*—Lobes, lobules, sulci, gyri on the surface of the hemispheres. Insular convolutions. Varieties and anomalies of the convolutions.

Vascular Supply of the Cerebro-Spinal Axis.

1. Spinal arteries, etc. Spinal branches, course, distribution, etc. Anterior and posterior spinal arteries. Nutritive arteries of the cord. Distribution in both gray and white matters.

2. Bulbar, pontine and cerebellar arteries. Vertebral, basilar, etc. Nutritive vessels of bulb and pons and ramifications and distribution as seen on section of neural tissues. Vascular supply of the cerebellum.

3. Arteries of the cerebrum. Main branches, distribution and ramifications. The cortical system and the ganglionic system.

Relations of Different Parts of the Encephalo-Spino-Neural System to the Skeleton.

Relation of cord and nerves to the bony column of the spine. Relations of cerebral gyri and sulci to the skull. Topography of the external surface of the skull. Definite landmarks of the skull's external surface. Primary and secondary areas of the skull. Contents of the respective areas. Topographical relations of the periphery of the body to the subdivisions of the cerebro-spinal axis.

Histological Elements of the Central Nervous System.

A. NERVOUS CONSTITUENTS: I, FIBRES; 2, CELLS.

B. NON-NERVOUS CONSTITUENTS: I, VESSELS; 2, EPITHELIA; 3, SUPPORTING TISSUES, TO-WIT: (A) MESOBLASTIC CONNECTIVE TISSUE, (B) EPIBLASTIC CONNECTIVE TISSUE.

A. Nervous constituents—

1. Nerve fibres—Various kinds found; minute structure; axis cylinder; primitive fibrillæ; spongioplasm, hyaloplasm; Fromann's stripes; axilemma; medullary sheath; myelin forms; cones of Lanterman; incisures of Schmidt; MacGowan's rods; nodes of Ranvier; myelemma; neurilemma; nuclei; nerve corpuscles; demilunes of Adamkiewicz. Histological subdivision of fibres: Primitive fibrillæ; naked axis cylinders; axis cylinders surrounded by neurilemma; axis cylinders surrounded by medullary tube; medullated fibres with neurilemma. Histological variations and developement

of medulla: Pathological changes; degenerations in medullated and non-medullated fibres; partial hypertrophy of cylinders; calcification.

2. Nerve cells. Relation to fibres; how to define. Varieties. Histology. Behavior to reagents. Isolation of cells. Characteristics of axial cylinder processes. Living cells. Granules. Homologies of granules. Nerve corpuscles. Histological meaning of nerve cells. Differences between motor and sensory cells. Golgi's views. Forel's views. Ehrlich's observations. Pathological changes. Atrophy. Fatty degeneration. Fatty pigmentary degeneration. Granular degeneration. Vacuole degeneration. Colloid change. Depigmentation. Hypertrophy. Calcification. Nuclear division. Cell division. Observations of Robinson, Ceccherelli, Fleischl.

B. Non-nervous constituents—

1. Vessels. Structure. Arteries, endothelium, membrana fenestrata. Coats. Limiting membrane. Adventitial lymph space. Space of Virchow-Robin; of His. Differences. Pericellular spaces. Veins and capillaries. Fat and pigment. Pathological changes in small vessels of the brain. Melanin. Fatty degeneration and calcification of vessels. Hypertrophy of walls. Pseudo-hypertrophy. Aneurysms. Corpuscles in adventitial spaces. Neoplastic elements. Coagula.

2. Epithelium—Difficulty of study in the adult human central nervous system. Study in the lower animals. In human animal best studied in central canal of the cord, floor of the fourth ventricle and aqueduct of Sylvius. Ependyma. Observations of Herms.

3. Supporting tissues: (a) Mesoblastic connective tissue; how to demonstrate; network of cells; fibres in the ependyma; pathology of this form of connective tissue; sclerosis; inflammatory processes, etc.

(*h*) Epiblastic connective tissue or neurogleia. Intercellular substance. Boll's interfibrillar granules, etc.

4. Other tissue elements. Fat granule cells. Amyloid bodies. Observations of Rindfleisch. Leber's corpuscles of Vincenti. Morbific micro-organisms. Neoplastic elements.

Peripheral Connections and Terminations of Nerves.

In the skeletal muscular system. In smooth muscle. In the cardiac musculature. In glandular structures, etc. Terminations of sensory nerves. Touch corpuscles of Wagner and Meissner. Vater's or Pacini's corpuscles. Krause's end bulbs. Genital corpuscles. Articulation nerve corpuscles. Tactile or touch corpuscles of Merkel, or corpuscles of Grandry. Herbst's corpuscles. Other modes of termination, as in the cornea, etc., in hair follicles, in tendons. Structure of the organ of smell. Regio olfactoria. Olfactory bulb, etc., Jacobsen's organ. Peripheral terminations of optic; of auditory nerves, etc.

Minute Structure of the Spinal Cord.

Structural features common to all the central nervous organs. Elements found, ground tissue, general arrangement of the cells and fibres. Causes of difference in color of the nervous structures. Subdivisions of gray matter in the central nervous system, to-wit:

1, the brain-mantle covering the surface of the secondary fore-brain; 2, rind of the cerebellum; 3, gray masses of the cord and corresponding portions of the brain from whence cranial nerves arise; inner lining of the third ventricle; 4, central ganglia. Topographical subdivision of white matter: 1, homodesmotic fibres; 2, heterodesmotic fibres. Nerve tracts in cord, comparison. Simplest scheme of connections of nerves and

nuclei. Explanation of term "nerve root." Nuclei of nerves which function symmetrically. Connections of nuclei which function symmetrically. Connections of nuclei with one another. Homologies of nuclei.

Histology of the Cord.

Of the white sheath. Periphery. Sustentacular framework. Direction of fibres. Size of fibres. Groups of visceral fibres. Distinction in gray matter. Arrangement of root fibres. Size and physical characteristics of cells in different locations in gray matter. Substantia spongiosa. Substantia gelatinosa. Its development. Central canal. Commissures of the cord.

Topography of the Cord.

Views on section, fissures, commissures, white and gray matter, cornua, etc. Arrangement of the cells. Anterior and posterior roots. Sections at different levels. Areas in cross sections. Course of the fibres in the cord. Relations of the posterior roots to the cord. Connections of the anterior roots. Origin of the posterior roots. Fibres which join those to Lockhart Clark's column. Do posterior roots join cells or plexus? Funiculi in white matter. Short and long tracts. Myelination of white matter. Connections of pyramidal columns. Statistics of degeneration. Cerebral visceral fibres. Degeneration of Goll's column. Fibres of the anterior commissure. Histogeny of the white columns.

Topographical Examination of the Brain.

Horizontal and vertical sections. Meynert's axis of section. Continuation of white and gray matter of the cord into bulbar regions. Changes produced by the pyramids. Displacement of the gray matter. Lateral columns in the bulb. Crossing of the fillet. Clava, etc. The

olives and the fibræ arcuatæ. Opening up of the central canal of the cord into the sinus rhomboidalis. Floor of the fourth ventricle. Posterior longitudinal bundle. Fillet. Roots of the VIIIth pair. Section of the pons. Roots of the abducens. Corpus trapezoides. Three limbs of the facial root. Large cells in the tegmental region. Roots of the VIth pair. Substantia ferruginea. Gray matter of the cerebellum. Origin of the trochlearis. Nucleus of this pair. Vestiges of ganglia in the IIIrd and IVth. Substantia nigra. Description of area viewed in section through the mid-brain. Front of the mid-brain. Nuclei of the IIIrd. Corpora geniculata, opening of the iter a tertii ad quartum ventriculum, and its relations to the cavity of the 'tween-brain. Sections of the great-brain. Relations of the basal ganglia. Nuclei of the 'tween-brain ganglia. Corpus subthalamicum. Ansa peduncularis. Front of the 'tween-brain.

Course of the Fibres of the Encephalo-Spino-Neural System, with their Relations, through the Bulbar and Pontine Planes, and in the Cord, Caudalwards and Brainwards.

Longitudinal fibres of the pedal system. Scheme of the pyramidal tracts. Proportion of fibres contained in them. Crural fibres for the cranial nerves. Constituents of the crusta. Fibres of the pontine planes. Transition of the crusta into the internal capsule. Connections of the posterior roots with the brain. Longitudinal fibres of the tegmental system. Nomenclature of the fillet. Constituents and degeneration of the fillet. Corpus restiforme and constituents. Fibres connected with the olives Connections of the lateral cerebellar tract. Upper termination of Gower's tract. The rest of the anterior and lateral columns. Posterior longitudinal fasciculus. Decussation of fibres in the mid-brain.

(16)

Schemes or Projections of the Nervous System.

That of Luys, of Meynert, of Alex. Hill, of Aeby, of Flechsig.

The Cranial Nerves.

I.—*Nervus olfactorius.* Comparison of this apparatus in man and lesser animals. Olfactory bulb. Its ventricle and minute anatomy. Homology of retina, olfactory bulb and spinal ganglia, nuclei, granules and ganglion cells. Olfactory tract, relation to anterior commissure. Connections of the tract: Gudden's experiments. Cortical centres of olfaction. Comparative anatomy. Relation of the temporal lobe to olfaction. General consideration of the great limbic arch in this connection. Function of the first pair.

II.—*Nervus opticus.* Divisions of the optic tract. Chiasm. Course of the fibres. Mesial root, etc. Fibres in the chiasm. The commissure of Gudden. Cortical visual areas. The corpora geniculata. Histology of the quadrigeminal bodies. Connections of the nates. Functions of the IInd pair.

III.—*Nervus oculomotorius.* Origin of root bundles for eye muscles. Nucleus. Course of fibres, their relation to the nucleus. Association of eye-muscle nuclei. Cortical centres. Peripheral distribution. Functions of the IIIrd pair.

IV.—*Nervus trochlearis*—Origin, course and distribution of fibres. Connections with other nerves. Functions of the IVth pair.

V.—*Nervus trigeminus*—Central origin. Convolutio trigemini. Course and exit from pons. Connections, portio-major, etc. Ascending root. Lateral addition. Median fibres. Descending root. Middle root. Cortical connections. Peripheral distribution. Physiological properties. Effects of division. Painful affections. Influence on the special senses.

XI.— *Nervus Accessorius*—Spinal and bulbar portions. Origin, course and distribution of fibres. Connections. Physiological properties.

XII.—*Nervus hypoglossalis*—Origin. Nucleus. Course and distribution of fibres. Passage through the formatio reticularis. Relation to the inferior olivary system. Function of the XIIth pair.

Pathological observations in connection with the preceding subject matter. Diseases of the brain-stem.

The Cerebellum.

Central nuclei. Medullary substance. Components of the brachium conjunctivum. Fibres in cerebellum. Commissures and connections. Elements of the cortex. Connections and central processes of the cellular elements of Purkinje. Distribution of the protoplasmic processes. Plane in which processes branch. Ultimate fate of processes. Tangential striation of molecular layer. Distribution of layers in lower vertebrates. Histogeny of cerebellum. Plan of the cerebellar cortex. Pathological observations in their relations to structural arrangement.

The Cerebrum.

Intermediate gray matter of the crural system. Intimate structure and connections and relations with other parts of the brain.

Thalamus opticus—Structure. Middle commissure. Fasciculus retroflexus. Pedunculi conarii. Thalamic connections. The peduncular systems.

Corpora Striata—Structure. Connections. Ansa lenticularis. The nuclei considered individually. Substantia nigra Soemmeringi.

The medullary centres of the great-brain. Classification of fibres in the centrum semi-ovale. Corona radiata.

The fornix. Corpora mammillare. Commissural fibres of the cerebrum. The corpus callosum, its meaning. The anterior commissure. Association fibres; fasciculus uncinatus, fasciculus longitudinalis inferior. Cingulum. Perpendicular occipital fasciculus of Wernicke.

Cortex cerebri—Intimate structure and physical characteristics. Layers. Gennari's stripe. Baillarger's lines. Pyramidal cells. Processes. Methods for showing fibres in the cortex. Meynert's five layers. Plan of cortex structure. Histological areas of the cortex. Special subdivisions of the cortex. Margin. Imperfectly developed parts. Cornu ammonis, its structure, etc. The fascia dentata and relation to the rind. Histogeny of the cortex.

The pineal body—Choroid plexus forms roof of the inter-brain. Farther back it becomes elongated into a tube, which is directed forward—the tube of the epiphysis. In some selacians, and in many reptiles, this passes through an opening in the skull to an organ of special sense, which strikingly resembles an eye. A cornea and a lens can be recognized, a retina and a pigmentary layer lying in and under the latter, in this "parietal eye." Discovered by Graaf and Spencer. In other vertebrates no relation can be detected between the epiphyseal tube and the organ of special sense, in the adult animal. The tube has disappeared in the depths of the skull, and the parietal eye, as is shown by transitional forms in amphibians and reptiles, is so completely lost that no trace of it can be found in birds or mammals. The blunt, knotted end of the tube remains as a nodule, the pineal gland, in front of the mid-brain.

Hypophysis cerebri—Appendix to base of brain about size of a cherry. Two parts; a prolongation of the floor of the ventricle (lobus infundibuli), which is not proven to consist of nerve-substance; and a tuft of epithelial

tubules attached to the lobus infundibuli, and derived from the mucous membrane of the pharynx. Intimate structure. Observations of Flesch, Darkschewitsch and others.

Pathological processes in their relation to anatomical data.

Membranes or Envelopes of Brain and Cord.

Contained material. Liquor cerebro-spinalis. Composition. How formed. Uses.

1, Meninx fibrosa, or dura mater; structure and relations; perivascular spaces; pathological observations.

2, Meninx serosa, or arachnoid; structure and relations; pacchionian bodies; pathological observations.

3, Meninx vasculosa, or pia mater; structure and relations; pathological observations.

Relations of tela choroideæ and plexus choroideæ. Pathology. Special consideration of the anatomy and pathology of the vascular apparatus of the great-brain.

Rotation of the great-brain during process of development, and observations in comparative anatomy.

The Sympathetic Nerve considered Anatomically, Histologically and Physiologically.

General distribution. Structure, ganglia, etc. Observations of F. Byron-Robinson. Vaso-motor, vaso-dilator, viscero-motor, inhibitory nerves, etc. Relation of sympathetic fibres to spinal fibres. Schema. Expansion of visceral or splanchnic divisions of certain spinal nerves. Observations of Michael Foster. Splanchnic fibres outside of the sympathetic. "Vascular" nerves in and out of the sympathetic.

Functions and subdivisions: 1, independent; 2, dependent: (*a*) cervical and cranial subdivisions. (*b*) thoracic, abdominal and pelvic.

Comparative anatomy. Historical resume.

PART II.

Functional classification of nerves.

I.   *Centrifugal or efferent*—A, motor; B, secretory; C, trophic; D, inhibitory; E, thermic; F, electrical.

Gaskell's classification—

  A—Nerves to vascular muscles: 1, vaso-motor; 2, vaso-inhibitory.

  B—Nerves to visceral muscles: 1, viscero-motor; 2, viscero-inhibitory.

  C—Glandular nerves: anabolic and katabolic nerves.

Examples and pathological observations.

II.--*Centripetal or afferent nerves*—A, sensory; B, special sense; C, reflex-motor, reflex-secretory, reflex-inhibitory.

III.--*Intercentral nerves.*

Physiology of Nerves.

Excitability of nerves.   Exhibition of function.   Stimuli: mechanical, thermal, chemical, electrical, physiological. Effects of currents.   Tetanus.   Unequal excitability. Diminution of excitability, fatigue, recovery.   Degeneration.   Wallerian law.   Regeneration of nerves.   Suture of nerves.   Transplantation of and union of nervous tissue.   Ritter-Valli law.   Excitable points.   Death of a nerve.

Electro-physiology—Physics of the question.   Instruments. Electrical currents in resting muscle and nerve.   Skin

(22)

currents. Rheoscopic limb. Action currents. Retinal and eye currents. Electrotonus.

Theories of muscle and nerve currents—Molecular, or pre-existence theory. Difference, or alteration theory. Transmission of nervous impulses. Double conduction in nerves. Electrical nerves. Electrical charging of the body. Comparative. Electrical fishes, etc.

Historical—Observations of Richer, Walsh, J. Davy, Becquerel, Brechet, Matteuci, von Humboldt, de Sauvages, A. Galvani, Volta, Pfaff, Bunzen, du Bois Reymond, Ranvier, Marey, Sanderson, Gotch and Ewart, among others.

Functions of the Spinal Cord.

The spinal nerves, functions. Effects of division or ablation. Effects of irritation. The ganglia, etc.

The spinal cord—As a conductor; results of hemisection; of section; experiments of Ludvig and Woroschiloff, of Dalton, etc.; centrifugal and centripetal paths; the muscular sense.

The cord as a collection of centres, or an independent centre. Reflex action. Inhibition of reflex action. Spinal co-ordination. Observations on the lower animals. Robin's experiments on guillotined criminals. Muscle tonus. Vaso-motor tonus. Trophic centres. Observations on their function. Heat centres. Tabulation of reflexes. General conclusions. The question of consciousness. Pathological observations in their relations to physiological data.

Functions of the Bulb.

Motor and sensory tracts. Centres; deglutition, respiration, cardiac, vaso-motor, pressor and depressor nerves. Lesions of the medulla oblongata. General resume and conclusions.

Functions of the Mesencephalon and Cerebellum.

I. *Maintenance of equilibrium*—Three factors: 1, system of afferent nerves and organs; 2, Co-ordinating centre; 3, Efferent tracts in connection with the muscular apparatus in action. Erect posture. Goltz' balancing experiments. Huxley's observations. Illustrations, etc.

    The afferent apparatus consists of: A, Organs for the reception and transmission of tactile or common sensory impressions; influence of tactile impressions; Volkman's and Heydt's observations; pathology. B, Organs for the reception and transmission of visual impressions; influence of visual impressions; the blind; sight in connection with motor adjustments regulation; disturbances by unusual movements in the field of vision; Cyon's observations. C, Semicircular canals of the internal ear and the afferent nerves; influence of labyrinthine impressions; experiments of Flourens; division of the semicircular canals; lesions of the same; mechanism of labyrinthine impressions.

II. *Co-ordination of locomotion*—Factors concerned in co-ordination. Locomotor ataxia. General considerations and conclusions.

III. *Instinctive or emotional expression*—Resemblances between animals deprived of their cerebral hemispheres and anæsthetized human subjects. Psychological observations, etc.

Functions of the Corpora Quadrigemina.

Relations to the special senses. Effects of destructive lesions. Effects of irritative experimentation. Observations of David Ferrier. Remarks on the phenomena. General deductions.

Functions of the Cerebellum.

Experiments of Flourens, those of David Ferrier, etc.
Effects of destructive lesions. Symptoms of cerebellar
disease. Irritative and destructive lesions compared.
Localized lesions. Electrical excitability. Results of
applications of electrical stimuli in various animals.
Galvanization of the head. Mechanism of cerebellar
co-ordination. Relative development of the cerebellum.
Owen's observations on cerebellar development in the
cyclostome and plagiostome cartilaginous fishes; Myxine,
lamprey and shark, etc. Recovery from cerebellar
lesions. Afferent relations of the cerebellum; relations
to common sensory tracts. Relation to the labyrinth;
to the eyes. Efferent relations of the cerebellum; to the
cerebrum; visceral relations. Pathological observations,
case of Alexandrine Labrosse reported by Combette.
Shuttleworth's case of congenital cerebellar atrophy.

Functions of the Cerebrum.

Methods of investigation. Views of Flourens. Excitability
of the cortex. Methods of stimulation. Conduction of
electrical currents. Excitability of medullary fibres as
compared with that of the brain-mantle. Observations
of David Ferrier, John C. Dalton, etc.
Phenomena of electrical irritation of the cortex—1, experiments on monkeys, convolutions of the simian brain; 2,
experiments on dogs; 3, experiments on various other
animals. Electrical stimulation of the basal ganglia,
corpora striata and the optic thalami.

The Hemispheres Considered Physiologically.

A. *General functional value of the cortical rind.*
Physiological topography of the hemispheres.

B. *The sensory centres:*—1, the visual centres; 2, the auditory centres; 3, the olfactory centres; 4, the gustatory centres; 5, the tactile centres.

C. *The motor centres:*—Topography of the motor areas Psycho motor paralysis. Bilateral association. Functional compensation. Consideration of sensory relations. The sense of effort. Muscular discrimination. The frontal motor centres. Anatomical relations of the frontal centres.

Physiology of the Senses.

Olfaction, vision, audition, taste, sensory and tactile sensation. Varieties. Law of peripheral perception. Pressure points, pain spots, tickling spots. Sense of locality. Pressure sense. Temperature sense. Common sensation. Pain. Muscular sense. The reproductive instinct.

PART III.

The Hemispheres Considered Psychologically.

Brain and mind. Conditions of perception. Feelings and
emotions. Appetites and desires. Motives to volition.
Ideation. Conflict of motives. Acquisition of speech.
Aphasia. Word-blindness and word-deafness. Control
of ideation. Attention. Substrata of attention. Phenomena of cerebral activity. Phenomena, genesis,
growth, development and perturbations of sensibility.
Genesis and evolution of memory. Psycho-intellectual
activity. Evolution of sensorial impressions. The
judgment, etc. Lessons taught by pathology and
experimentation. Psychological definition of personality. Hypothesis of the ego. Genesis and development of the ego. Experimental psychology. The
hypotheses concerning the nature of consciousness.
Psychic and nervous activities. Psychometric researches.
Observations of Maudsley and Herzen, of Ribot.
Preyer's notes on brain metabolism in connection with
psychic activity. Sleep. Observations of Despine, of
Hartmann. Organic conditions of personality. The
principle of individuation. Cœnæsthesis. Remarks
of Louis Peisse, of Coudillac, etc. The basic feeling of
organic life. Its elements. Contributions on the part
of the various vital functions; organic sensations attached
to respiration, those from the alimentary canal, those
connected with the state of nutrition, general and local
circulation. Organic sensations arising from the state
of the muscles; associations with the special senses.
Organic sensations of the genital apparatus. Compre-
hension of the physical bases of personality. Their

relation to the higher forms of mental life. Extra-physiological variations and centripetal stimuli as ætiological factors in dream production. Does physical personality exist in nature? Comparison between the higher and lower forms of life. Variations of personality in the normal state; euphory. Exaltation and depression. Borderland between the normal and the abnormal. Reductions constituting pathological states. Double personality. Personality of double montrosities; of twins. Emotions in their relation to personality. Metamorphosis of personality. Normal and abnormal sexual characters, hermaphrodites, eunuchs, sexual perverts. Pathological phenomena in their relation to psychological data. Dissolution of personality. Comparative psychology. Evolution of zoological individuality. Hypnotism, etc.